ANIMAL BATTLES

WOLVERINE VS. HONEY BADGER

TORQUE™

BY KIERAN DOWNS

BELLWETHER MEDIA • MINNEAPOLIS, MN

Torque brims with excitement perfect for thrill-seekers of all kinds. Discover daring survival skills, explore uncharted worlds, and marvel at mighty engines and extreme sports. In *Torque* books, anything can happen. Are you ready?

This edition first published in 2021 by Bellwether Media, Inc.

Library of Congress Cataloging-in-Publication Data

Names: Downs, Kieran, author.
Title: Wolverine vs. honey badger / by Kieran Downs.
Other titles: Wolverine versus honey badger
Description: Minneapolis, MN : Bellwether Media, 2021. | Series: Torque: animal battles | Includes bibliographical references and index. | Audience: Ages 7-12 | Audience: Grades 4-6 | Summary: "Amazing photography accompanies engaging information about the fighting abilities of wolverines and honey badgers. The combination of high-interest subject matter and light text is intended for students in grades 3 through 7"– Provided by publisher.
Identifiers: LCCN 2020041126 (print) | LCCN 2020041127 (ebook) | ISBN 9781644874639 (library binding) | ISBN 9781648342561 (paperback) | ISBN 9781648341403 (ebook)
Subjects: LCSH: Wolverine–Juvenile literature. | Honey badger–Juvenile literature.
Classification: LCC QL737.C25 D69 2021 (print) | LCC QL737.C25 (ebook) | DDC 599.76/6–dc23
LC record available at https://lccn.loc.gov/2020041126
LC ebook record available at https://lccn.loc.gov/2020041127

Editor: Christina Leaf Designer: Josh Brink

Printed in the United States of America, North Mankato, MN.

TABLE OF CONTENTS

THE COMPETITORS

WOLVERINE

Two tough **mammals** are ready to fight! Both the wolverine and the honey badger are known for their long claws, sharp teeth, and fearless **attitudes**.

These animals are feared in their home areas. But what would happen if they crossed paths? Who would win in a battle of the brave?

HONEY BADGER

Wolverines roam northern forests. They have short legs and long bodies. Their brown fur has a light brown stripe that goes down both sides of their bodies.

Wolverines do most of their hunting at night. They also **scavenge** a lot of their food. Wolverines will eat just about any food that they find!

LARGE WEASELS

Wolverines may look like bears, but they are actually a part of the weasel family

WOLVERINE PROFILE

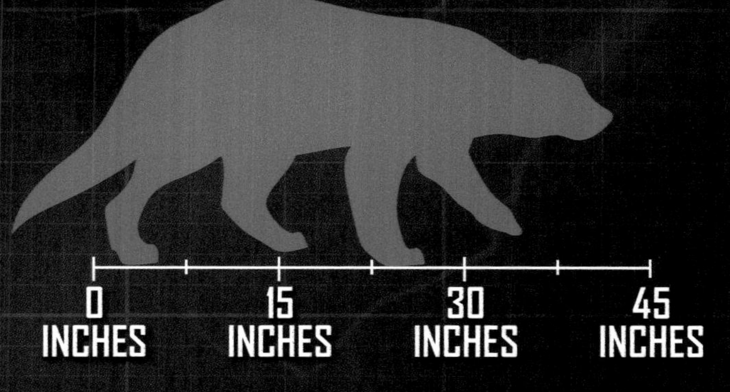

LENGTH

UP TO 41 INCHES
(104 CENTIMETERS)

WEIGHT

UP TO 66 POUNDS
(30 KILOGRAMS)

HABITAT

MOUNTAINS

GRASSLANDS

FORESTS

TUNDRA

WOLVERINE RANGE

HONEY BADGER PROFILE

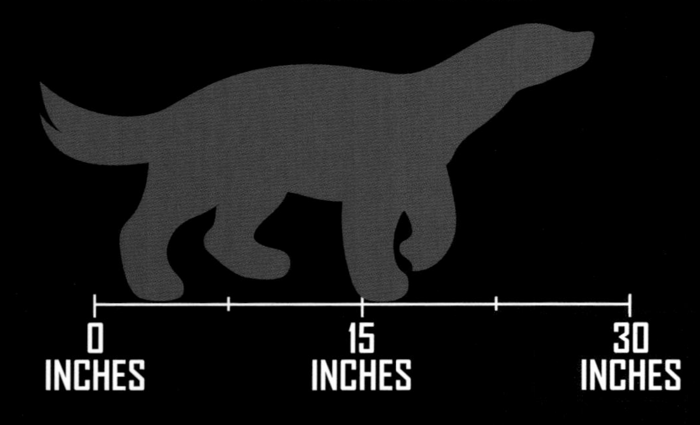

LENGTH

28 INCHES
(71 CENTIMETERS)

WEIGHT

UP TO 30 POUNDS
(14 KILOGRAMS)

HABITAT

FORESTS

DESERTS

MOUNTAINS

GRASSLANDS

HONEY BADGER RANGE

Honey badgers are also called ratels. Their dark, flat bodies have a white stripe across the top. Honey badgers are **omnivores**. They earned their name because they seek out beehives. They like to eat the bee **larvae** and honey inside.

Honey badgers live in a variety of **habitats**. They can be found in warm and cold **climates**.

SNAKE ON THE MENU

Honey badgers will eat snakes. They are not afraid of getting bitten. Snake poison does not harm them!

SECRET WEAPONS

Wolverines have five long, sharp claws on each paw. These are used to catch **prey** and fight enemies. They are also helpful for climbing and digging.

Honey badgers also have long front claws to capture prey. They can reach up to 1.5 inches (4 centimeters) long! Their sharp teeth are also useful in taking down enemies.

WOLVERINE

SHARP CLAWS

STRONG TEETH

SENSE OF SMELL

Wolverine teeth are strong and sharp. The back teeth are in the mouth at an angle. This lets wolverines bite into frozen meat. They can even eat bones!

SECRET WEAPONS

HONEY BADGER

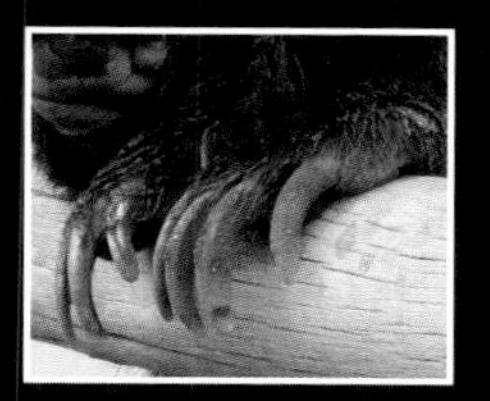

LONG CLAWS

STINK

LOOSE SKIN

Honey badgers store a stinky liquid by their tail. They spread the stink when they feel threatened. This sends most enemies running.

Wolverines have a strong sense of smell. They can sniff out prey under up to 20 feet (6 meters) of snow. They dig underground to attack **hibernating** prey.

Honey badgers have loose, tough skin. If an enemy grabs them, they can still twist and turn to fight. The skin also lets them take a lot of hits from enemies.

ATTACK MOVES

Honey badgers are often very **aggressive**. They are not afraid to fight almost any enemy. They attack with their sharp teeth and claws.

Wolverines can hunt prey much bigger than themselves. They do best in winter. They chase prey until it gets stuck in deep snow. Then they attack!

SNOWSHOES

Wolverine paws spread out to almost twice as wide when walking on snow. This makes it easier to stay on top of the snow!

When wolverines catch their prey, they attack with their teeth and claws. They bite their enemies at the back of the neck to take them down.

Honey badgers are able to fight for a long time. Enemies often cannot get through their tough skin. The badgers will continue to attack until their enemies are too tired to fight anymore.

READY, FIGHT!

A wolverine smells food. But a honey badger is already there. The fight is on!

The wolverine attacks the badger with its sharp claws. But the honey badger does not care! It fights back with its own claws. The wolverine cannot break through the honey badger's thick skin. It runs off. The badger gets to keep its meal!

NO FEAR

The honey badger holds the Guinness World Record for being the most fearless animal in the world!

GLOSSARY

aggressive—ready to fight

attitudes—ways of thinking or feeling that affect behavior

climates—long-term weather conditions for certain areas

habitats—homes or areas where animals prefer to live

hibernating—spending the winter sleeping or resting

larvae—young insects; larvae often look like worms.

mammals—warm-blooded animals that have backbones and feed their young milk

omnivores—animals that eat both plants and animals

prey—animals that are hunted by other animals for food

scavenge—to look for food to eat that is already dead

TO LEARN MORE

AT THE LIBRARY

Pallotta, Jerry. *Hyena vs. Honey Badger.* New York, N.Y.: Scholastic Inc., 2018.

Stewart, Melissa. *Wolverines.* Washington, D.C.: National Geographic Kids, 2018.

Wilsdon, Christina. *Ultimate Predatorpedia: The Most Complete Predator Reference Ever.* Washington, D.C.: National Geographic Kids, 2018.

ON THE WEB

FACTSURFER

Factsurfer.com gives you a safe, fun way to find more information.

1. Go to www.factsurfer.com
2. Enter "wolverine vs. honey badger" into the search box and click 🔍.
3. Select your book cover to see a list of related content.

INDEX

The images in this book are reproduced through the courtesy of: DenisaPro, cover (wolverine); laurenpretorius / Getty Images, cover (honey badger); Bogdanov Oleg, p. 4; Lauren Pretorius, p. 5; belizar, pp. 6-7; Phillip Bird / Alamy Stock Photo, pp. 8-9; Michal Ninger, pp. 10, 12 (sense of smell, sharp claws, strong teeth); Erwin Niemand, p. 11; TashaBubo, pp. 12, 20-21 (wolverine head); Braam Collins, p. 13; Nora Mari, p. 13 (long claws); Michael Potter11, p. 13 (stink); Matt Gibson, p. 13 (sharp teeth); CREATISTA, p. 14; Ann & Steve Toon, p. 15; Vincent Grafhorst, p. 16; Robert Postma / Alamy Stock Photo, p. 17; User10095428_393, p. 18; Dirk.D.Theron, p. 19; AfriPics.com / Alamy Stock Photo, pp. 20-21 (honey badger); chris willemsen / Alamy Stock Photo, pp. 20-21 (honey badger head); Richard Seeley, pp. 20-21 (wolverine).